WHAT DOES IT LOOK LIKE?

IT HAS WINGS!

BY KATHLEEN CONNORS

Gareth Stevens PUBLISHING

first concepts

Many animals
have wings.
Wings help animals fly.

Birds have wings.

Robins are birds.

They have wings.

Cardinals have wings.

Ducks have wings.

Eagles have wings.

Many bugs have wings.
Butterflies have wings.

Bees have wings.

Ladybugs have wings.

Bats are the only mammals that fly. They have wings!

An airplane
has wings too!

Which animal has wings?

Please visit our website, www.garethstevens.com. For a free color catalog of all our high-quality books, call toll free 1-800-542-2595 or fax 1-877-542-2596.

Cataloging-in-Publication Data

Names: Connors, Kathleen.
Title: It has wings! / Kathleen Connors.
Description: Buffalo, NY : Gareth Stevens Publishing, 2026. | Series: What does it look like?
Identifiers: ISBN 9781482470963 (pbk.) | ISBN 9781482470970 (library bound) | ISBN 9781482470987 (ebook)
Subjects: LCSH: Wings (Anatomy)–Juvenile literature.
Classification: LCC QL950.8 C747 2026 | DDC 591.47′9–dc23

Published in 2026 by
Gareth Stevens Publishing
2544 Clinton Street
Buffalo, NY 14224

Designer: Andrea Davison-Bartolotta
Editor: Kristen Nelson

Photo credits: Cover, p. 1(background) Honyojima/Shutterstock.com; cover, pp. 1 (teddy bear), 23 (dog) Eric Isselee/shutterstock.com; p. 3 (bat) Passakorn Umpornmaha/Shutterstock.com; p. 3 (bird) Element allqulity Designe/Shutterstock.com; p. 3 (fly) Vinicius R. Souza/Shutterstock.com; p. 3 (moth) Sandra Standbridge/Shutterstock.com; p. 5 Stephen A. Waycott/Shutterstock.com; p. 7 Craig Sterken/Shutterstock.com; p. 9 aabeele/Shutterstock.com; p. 11 Linda J Scher/Shutterstock.com; p. 13 Shabi2731/Shutterstock.com; p. 15 Fidel/Shutterstock.com; p. 17 Christian Mueller/Shutterstock.com; p. 19 Rudmer Zwerver/Shutterstock.com; p. 21 Thierry Weber/Shutterstock.com; p. 23 (moth) romm/Shutterstock.com; p. 23 (shark) frantisekhojdysz/Shutterstock.com; p. 23 (snake) Kurit afshen/Shutterstock.com; p. 23 (sugar glider) dwi putra stock/Shutterstock.com.

Printed in China

CPSIA compliance information: Batch #QSGS26: For further information contact Gareth Stevens at 1-800-542-2595.